First published in 2009 by
MACMILLAN EDUCATION AUSTRALIA PTY LTD
15–19 Claremont Street, South Yarra 3141

Visit our website at www.macmillan.com.au or go directly to www.macmillanlibrary.com.au

Associated companies and representatives throughout the world.

National Library of Australia Cataloguing-in-Publication entry

Richards, Julie.
  Solar energy / Julie Richards.
  9781420267174 (hbk.)
  Energy choices
  Includes index.
  For primary school age.
  Solar energy - Juvenile literature
   Renewable energy sources - Juvenile literature.
333.7923

Text and cover design by Christine Deering
Page layout by Domenic Lauricella
Photo research by Legend Images
Illustrations by Richard Morden

Printed in China

Acknowledgements
The author and the publisher are grateful to the following for permission to reproduce copyright material:

Front cover photograph: Solar power station, Australia courtesy of Photolibrary /John Mead/SPL

Photos courtesy of:
123rf/Leszek Scholz, 11 (top); AAP Image/Juan Ferreras, 26; AAP Image/Dean Lewins, 23; AAP Image/Photoalto, 3, 8; © Chiyacat/
Dreamstime.com, 21; © Christopher Elwell/Dreamstime.com, 18 (centre); © Kabby/Dreamstime.com, 25; © Qilux/Dreamstime.com,
18 (right); Denis Doyle/Getty Images, 16; Mel Yates/Getty Images, 30; © Lena Andersson/iStockphoto, 12; © Robert Churchill/
iStockphoto, 10 (left); © faberfoto_it/iStockphoto, 10 (right); © Nick Free/iStockphoto, 6; © Clayton Hansen/iStockphoto, 4 (top);
© Justin Horrocks/iStockphoto, 11 (bottom); © Daniel Stein/iStockphoto, 5; © Adam Tomasik/iStockphoto, 13; © Georg Winkens/
iStockphoto, 14; © Macquarie Generation, 17; NASA/JPL-Caltech, 24; NREL/DOE, photo by Mike Linenberger, 22; Photolibrary ©
Mark Boulton/Alamy, 28; Photolibrary © David Hancock/Alamy, 19; Photolibrary/Photo Researchers, 18 (top left); Photolibrary/Ralph
Reinhold, 18 (bottom left); Photolibrary /John Mead/SPL, 1; Photolibrary/Hank Morgan/SPL, 15; Sandia National Laboratory, 27;
Shutterstock, 29; © Petr Nad/Shutterstock, 20; © Jaimie D. Travis/Shutterstock, 4 (bottom).

While every care has been taken to trace and acknowledge copyright, the publisher tenders their apologies for any accidental
infringement where copyright has proved untraceable. Where the attempt has been unsuccessful, the publisher welcomes
information that would redress the situation.

Contents

Glossary words

When a word is printed in **bold**, you can look up its meaning in the Glossary on page 31.

What is energy?

Energy makes things work. Many machines need electrical energy to work. The more machines we use, the more electrical energy, or **electricity**, we need. Electrical energy is made in power stations.

a toaster

a laptop computer

These machines use electricity from a power station.

Most of the energy we use is made by burning **fossil fuels**, such as coal. Fossil fuels are running out because we use them too much. We need to use other sources to make **alternative energy**.

Burning fossil fuels releases pollution into the air.

Renewable energy

Energy sources that will not run out are called renewable sources. Energy sources that will run out are called non-renewable sources. Fossil fuels are a non-renewable energy source.

The Sun is a renewable energy source.

Sustainable energy

Sustainable energy is made from renewable energy sources. These sources will still be available in the future. They will not run out.

Comparing energy sources		
Energy source	Renewable	Sustainable
Solar energy	✔	✔
Wind energy	✔	✔
Water energy	✔	✔
Nuclear energy	✘	✘
Biofuels	✔	✔
Fossil fuels	✘	✘

Solar energy

Energy that comes from the Sun is called solar energy. Solar energy is a renewable energy source.

You can see solar energy as sunshine and feel it as heat.

The Sun makes solar energy. Although the Sun is far away, its energy travels quickly through space to Earth.

Each explosion sends light and heat into space.

Gases in the Sun cause huge explosions.

The light and heat travels in waves to Earth.

The Sun's energy brings light and heat to Earth.

Solar energy in nature

Natural solar energy warms the Earth and gives us sunlight. Natural solar energy:

- helps plants grow
- makes flowers open
- ripens fruits.

Plants need sunlight to survive.

Natural solar energy heats the Earth's deserts.

People use natural solar energy to dry their washing on a clothes line. Natural solar energy can be used in buildings and houses to help light and heat them.

Wet clothes dry quickly on a sunny day.

Skylights let light into rooms where there are no windows.

Making electricity

Solar energy can be turned into electricity. Solar cells **absorb** the Sun's energy and use it to make electricity. The electricity is used straight away or stored in a **battery**.

The sunnier it is, the more solar energy will be absorbed by this solar cell.

Many solar cells can be joined together to make a solar panel. Solar panels can be placed wherever they are needed.

Solar panels can be placed on the roof of a house.

Solar farms

Solar farms have thousands of solar panels tilted towards the Sun. The panels absorb energy from the Sun.

Solar farms need a lot of space for all the solar panels.

Some solar farms are built in deserts. Deserts have little or no rain, so there are fewer clouds to block the Sun's energy.

Deserts are a perfect environment for solar farms.

Solar power stations

Some power stations use a solar power tower.
Mirrors reflect and focus sunlight onto the tower.
The energy is collected to make electricity.

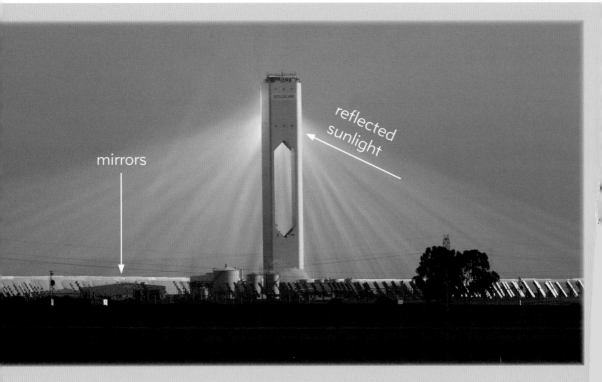

mirrors

reflected sunlight

This solar power tower in Spain uses 624 mirrors
to reflect sunlight.

Sometimes solar power stations are built next to coal-fired power stations. When it is sunny, the solar power station can make the electricity.

On cloudy days, the coal-fired power station makes the electricity.

Using solar energy

Solar energy can be used to power many things at home. Calculators, torches, radios and even toys can use solar energy.

a solar-powered calculator

a solar-powered radio

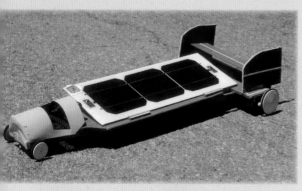

a solar-powered toy car

a solar-powered battery charger

There are many solar-powered things that can be used at home.

Solar energy can be used to heat swimming pools and greenhouses. Solar lights can be used near garden paths and driveways.

This street light absorbs and stores solar energy during the day and lights up at night.

Solar energy at home

Solar energy can be used to heat water at home. The water runs slowly through the pipes inside a solar water heater and heats up. The hot water is stored in a tank for showers, baths and washing.

Rooftop solar panels are used to heat water for this house.

Solar energy can also be used to make electricity in the home. When the Sun is not shining, electricity from other sources is used instead.

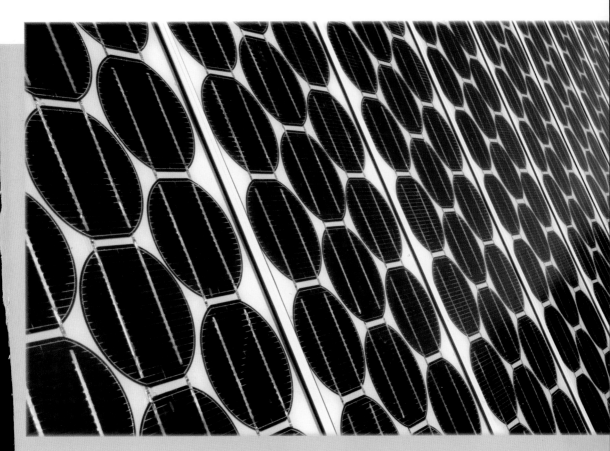

These rooftop solar panels, made up of many solar cells, turn sunshine into electricity.

Solar-powered transport

Solar energy can be used for transport. Solar cells can power small cars that do not need to travel very far.

solar panels

This car can recharge its battery using solar energy.

Solar-powered boats have solar panels to absorb the Sun's energy from the sky. The panels also absorb energy from sunlight reflected off the water.

Solar panels power this passenger ferry.

Solar energy in remote places

Solar energy powers equipment used in spacecraft and **satellites**. Remote-controlled robots are also powered by solar energy. They explore planets humans cannot travel to.

This solar-powered robot explored the planet Mars.

Solar energy is useful in mountainous areas or deserts. It is too expensive and difficult to build coal-fired power stations in these **remote** environments.

In remote places, solar energy can power equipment such as this solar light.

The future of solar energy

Scientists are working on ways to make solar energy cheap enough for everyone. This will help **conserve** the world's energy sources.

Solar dishes like this can be used for cooking.

A large solar farm is being built in the desert in the United States. Solar dishes will reflect the Sun's energy onto engines to make electricity. This solar farm will be the world's largest solar energy system.

These solar dishes follow the Sun as it moves across the sky.

Using less energy

Using less energy will conserve energy sources and reduce pollution. Buying **appliances** that are **energy-efficient** is one way to use less energy in the home.

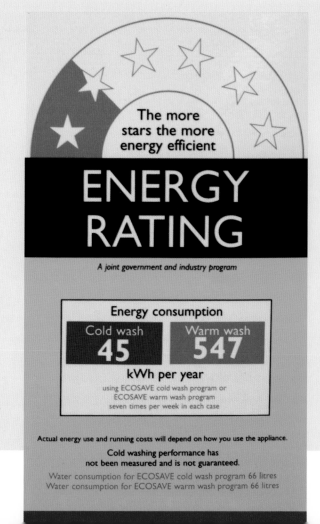

Energy rating labels tell you how energy-efficient an appliance is.

The more stars the more energy efficient

ENERGY RATING

A joint government and industry program

Energy consumption

Cold wash	Warm wash
45	547

kWh per year

using ECOSAVE cold wash program or ECOSAVE warm wash program seven times per week in each case

Actual energy use and running costs will depend on how you use the appliance.

Cold washing performance has not been measured and is not guaranteed.

Water consumption for ECOSAVE cold wash program 66 litres
Water consumption for ECOSAVE warm wash program 66 litres

Today, many things are made to use less electricity than in the past.

Low-energy light globes do not use as much electricity as traditional light globes.

How can we help?

Everybody can help the environment by using less energy. We can use less energy by:

- switching off lights and appliances when they are not being used

- asking for presents that do not use electricity to work.

If you feel cold, put on a jumper instead of turning up the heater.

Glossary

absorb	to soak up
alternative energy	energy made from a source such as the Sun instead of fossil fuels
appliances	machines such as televisions or computers that need electricity to work
battery	a container that stores energy
biofuels	fuels made from plant and animal matter
conserve	not waste
electricity	electrical energy that is carried along wires
energy-efficient	uses energy without waste
fossil fuels	coal, oil and gas
nuclear energy	energy stored inside the centre of an atom
remote	a long way away from cities and diffcult to get to
satellites	spacecraft that circle the Earth

Index